The Operation of Autonomous Underwater Vehicles

Volume One:

Recommended Code of Practice for the Operation of Autonomous Marine Vehicles, Second Edition

The papers published within reflect the opinions of the individual authors and not the Society for Underwater Technology or the other sponsors unless specifically stated. The Society does not accept responsibility for the technical accuracy or any items printed in *The Operation of Autonomous Underwater Vehicles, Volume One: Recommended Code of Practice for the Operation of Autonomous Marine Vehicles, Second Edition.*

Front cover image
A HUGIN 1000 vehicle on a Royal Norwegian Navy mine hunter during a deployment in the standing NATO MCM force
Courtesy of Kongsberg Maritime, AUV R&D Department

© Society for Underwater Technology 2000, 2009
80 Coleman Street, London EC2R 5BJ, UK
All rights reserved. First edition 2000
Second edition 2009

ISBN 0 906940 51 6
ISBN-13 978 0 906940 51 8

Citation
Society for Underwater Technology. (2007). *Volume One: Recommended Code of Practice for the Operation of Autonomous Marine Vehicles, Second Edition.* The Operation of Autonomous Underwater Vehicles. London: Society for Underwater Technology, 78pp.

A CIP catalogue record for this book is available from the British Library.

Designed by Mariam Pourshoushtari, Publications Officer, SUT
Printed by Primary Colours and bound by the Blissett Group, London, UK

This publication is protected by international copyright law. No part of this publication may be reproduced or utilised in any form or by any means, electronic or mechanical including photocopying, recording or by any information storage and retrieval system, without the prior permission of the publishers.

Contents

Foreword	7
Introduction	11
a. Application	11
b. Background	11
c. AMV Characteristics	12
d. AMV Classification	13
e. AMV Missions and Tasks	13
Code of Practice	15
1. Legal Considerations	15
1.1. Compliance with Existing Framework	15
1.2. Periodic Review	15
1.3. Precautionary Principle	15
2. Contractual Considerations	16
2.1. Capabilities and Limitations	16
2.2. Delivery of Products	16
2.3. Fallback	16
3. Environmental Considerations	16
3.1. Pollution	16
3.2. Vehicle Identification	17
3.3. Sonar, Lasers, and Other Electromagnetic Emitters	17
3.4. Chemicals	17
3.5. Mechanical Interaction with Third Parties	18
3.6. Environmental Impact Assessment	18
4. Operational Considerations	18
4.1. Planning	19
4.1.1. Access to Operating Area	19
4.1.2. Operating Site Suitability	19

	4.1.3.	Installation Drawing	19
	4.1.4.	Custom Adapters and Fittings	20
	4.1.5.	Communications	20
4.2.	*Risk Assessment and Management*		20
	4.2.1.	Forms of Risk Management	20
	4.2.2.	Review and Update	21
	4.2.3.	Feedback and Effects	22
	4.2.4.	AMV System Location and Integrity	22
4.3.	*Influence of the Environment*		24
	4.3.1.	Operating Envelope	24
	4.3.2.	Analysis and Forecasts	24
	4.3.3.	Water Depth	24
	4.3.4.	Temperature and Weather	25
	4.3.5.	Water Density (Salinity)	25
	4.3.6.	Pollutants	26
	4.3.7.	Currents	26
	4.3.8.	Sea State and Swell	27
	4.3.9.	Seabed Characteristics	28
4.4.	*Operating Procedure*		27
4.5.	*Handling Systems*		28
	4.5.1.	Introduction	28
	4.5.2.	Documented Procedures	28
	4.5.3.	Sea State	29
	4.5.4.	Supervision and Communications	29
	4.5.5.	Maintenance	29
4.6.	*Pre- and Post-Mission Procedures and Checks*		29
4.7.	*Communications and Relocation*		30
	4.7.1.	Team Briefing	30
	4.7.2.	Radio Communications	
	4.7.3.	Vehicle Relocation	30

		4.7.4. Neighbouring Activities	31
		4.7.5. Communications with the AMV	31
	4.8.	*Navigation*	31
	4.9.	*Manuals and Documentation*	32
	4.10.	*Hazards to Operating Personnel*	32
		4.10.1. Safe Use of Electricity	32
		4.10.2. Moving Parts	33
		4.10.3. Lasers	33
		4.10.4. Use of Divers	33
		4.10.5. Manual Handling	34
5.	**Equipment Certification and Maintenance**		34
	5.1.	*Planned and Periodic Maintenance*	34
		5.1.1. Design, Testing, and Periodic Examination of Handling Systems	35
	5.2.	*Spare Parts*	36
	5.3.	*Equipment Register*	36
6.	**Personnel**		36
	6.1.	*Qualifications and Experience*	36
	6.2.	*Team Size*	37
	6.3.	*Working Periods*	39
	6.4.	*Training*	39
		6.4.1. Safety Training	39
		6.4.2. Technical Training	39
	6.5.	*Language*	40
	6.6.	*Logbooks*	40
	6.7.	*Allocation of Responsibilities*	40
		6.7.1. Team Leader	41
		6.7.2. Team Members	42
		6.7.3. Other Personnel	42
References			44
Glossary of Terms			45

ANNEX A: DEGREES OF AUTONOMY   47

ANNEX B: CLASSIFICATION FOR RISK MANAGEMENT PURPOSES   49

ANNEX C: EXAMPLE TASKS   51
    C.1. Oceanic Process Studies   51
    C.2. Routine Observations   52
    C.3. Survey   53
    C.4. Inspection   53

ANNEX D: AMV TOOLS   55
    D.1. Power Constraints   55
    D.2. Marine Science AMV Tools   55
    D.3. *In Situ* Sensors   56
    D.4. Sampling Devices   56
    D.5. Remote Sensing Devices   56
    D.6. Industrial AMV Tools   57
    D.7. Military AMV Tools   57

ANNEX E: MEMBERSHIP OF THE SUT AUTONOMOUS UNDERWATER VEHICLE LEGAL WORKING GROUP (AUVLWG) 2009   59

APPENDIX 1: STATUS OF THE LAW/CURRENT LEGAL ISSUES REGARDING AMVs   61

APPENDIX 2: IMO – REVISED TECHNICAL ANNEX II TO THE DRAFT CONVENTION ON THE LEGAL STATUS OCEAN DATA ACQUISITION SYSTEMS (ODAS)   75

Foreword

This update to the first edition of the Code of Practice (CoP) for autonomous underwater vehicles (AUV) has been produced by the successor to the Society for Underwater Technology's Autonomous Underwater Vehicles Legal Working Group (AUVLWG) to provide a pertinent reference document for the safe and efficient operation of autonomous marine vehicles (AMV). It has been widened to include any form of autonomous vehicle – as exemplified by the term AMV – operating in the marine environment either on or under the water that is acting in a fully autonomous mode. It does not apply to autonomous surface vehicles (ASV) which are large and/or fast. Annex E gives details of the membership of the AUVLWG.

The original CoP drew heavily on the International Maritime Contractors Association (IMCA)[1] *Code of practice for the safe and efficient operation of remotely operated vehicles*. This update is now based on several years of experience in operating these vehicles.

It is of particular concern now that AMVs are brought into the marine legal regime, as they lie outside it for most of their operations. Consultation on the path to legal recognition of AMVs has led to the following points, many of which are covered within this document:

a. Definition – covered in Section 1 and Annex A

b. Rules for operation, to include a Code of Conduct for Operators

c. Rules for marking (including lighting) – a draft set of guidelines on markings, lights, and signals for ocean data acquisition systems (ODAS) was proposed by the International Maritime Organisation's (IMO's) Maritime Safety Committee in 1984 (Appendix 2); the general guidance is relevant to AMVs.

Other points will form additions to this document at a later date:

a. Rules for water space management – new section
b. Rules for operating in the vicinity of deployment and third parties – it is anticipated that this will be part of the section on water space management
c. Rules on discovering a unit, salvage, protection of data, etc. – new section; the principles and areas of uncertainty are described in Brown and Gaskell (2001)[2]
d. Conventions for pursuing operators in event of damage to third parties – new section.

This AMV CoP has been written as a voluntary code that the SUT endorses for adoption by the AMV community. It is a non-legal document which encapsulates the combined experience of the members of the AUVLWG, spanning most – if not all – aspects of civil and military AMVs and should be regarded as a guide, based on best practice, to the issues to be considered in the design, build, and operation of an AMV. The code is primarily aimed at the users, designers, researchers, and manufacturers of AMV systems in the United Kingdom. Though the CoP is focused on the UK, it is believed by both the AUVLWG and the SUT that it could be used as a basis of a codified procedure to be adopted by the international AMV community.

The code does not remove any existing legal obligation from the AMV community under current national EU or international law. Local or national regulations always take precedence over this code.

The SUT hopes that adoption of this CoP and adherence to the guidance in it will lead to a safe and efficient industry that operates to common standards.

This CoP should be treated as a dynamic document and the advice given in it will, by necessity, change as the industry continues to grow. It is intended to review this code on a regular basis and make any necessary improvements based on a consensus approach.

Suggestions for improvement may be forwarded to:

The Chief Executive

SUT Head Office   80 Coleman Street   London EC2R 5BJ   UK
t 020 7382 2601   **f** 020 7382 2684   **e** info@sut.org

This is a Code of Practice, not a commentary on national and international law.

Introduction

a. Application

This Code of Practice (CoP) is intended to apply only to marine vehicles being operated in the autonomous mode, either submerged or surfaced and not under direct control (both as defined in the body of the document). It does not cover any vehicle with a warhead (a destructive weapon of war), which is outside the scope of the document and the discussions contained within. It also does not cover large and/or fast autonomous surface vehicles (ASV).

b. Background

This SUT publication provides the United Kingdom AMV community with a Code of Practice for use and voluntary adoption as a community standard. The concept of AMVs has been in evidence for at least twenty years and has a wide range of operating capabilities and technical solutions. The safety and legal framework for their use, however, has yet to be properly formulated. The wide and everyday use of the technology is still relatively immature as an established part of the marine technology scene.

Early AMV development and research have their roots in the military maritime operational requirement, with major maritime powers such as the United States and Russia being examples of leading developers of an early military AMV capability. However, more recently in the UK, Norway, and elsewhere, both the civil research and industrial communities have developed ocean-going AMV capabilities of international repute.

AMV development and, more importantly, routine usage are increasing. This is partly being driven by the costs of ship time incurred when performing routine operations, such as pipeline inspection or repetitive oceanographic measurements.

This growth of AMV usage has occurred under a loose legal framework, which, because of the current legal status of the technology, has meant that the AMV community has had to be self-regulating. This code is in some way an expression of that self-regulatory approach.

c. AMV Characteristics

Much has been written[3,4] on the legal and policy issues related to AMV use. These writings have produced a definition for an AMV that, it is hoped, will become the community standard for an AMV operating in an autonomous mode, as given in this section.

The Autonomous Underwater Vehicles Legal Working Group (AUVLWG) submits that the defining characteristics of an AMV, operating in an autonomous mode are that it is:

(i) A man-made unmanned vehicle that operates in the marine* environment;

(ii) Not mechanically linked to, or restrained by, an operational/launch station (e.g. boat, submarine, shore station);

(iii) Operating untethered (e.g. no fibre optic or wire communications link) and with no form of direct control** (e.g. radio or acoustic, although it may also be capable of receiving and acting on further instructions);

(iv) Capable of movement, with a horizontal component, relative to surrounding water mass; and

(v) Capable of onboard decision-making.

It must be stressed that the description of a physical system as an AMV or an ROV is not necessarily a permanent property of the system and may vary with time. For example, a system may be designed to operate primarily under direct control, but can transfer to autonomous mode should the control link be broken. A vehicle may also be specifically designed to operate as an AMV during part(s) of its mission, and as an ROV at other times.

---

* Whilst this CoP describes systems in use in the marine environment, it may also be applied to operations in inland or other waters.

** Direct control is defined as follows: the vehicle responds immediately to an operator command, constrained only by communication and processing time. In the absence of any operator command, the vehicle travels at constant velocity.

The degree of autonomy or decision-making may vary from the simplest logic based on an environmental factor (e.g. a glider changes to positive buoyancy when it reaches a predetermined depth), through to the most advanced (for the current time) artificial intelligence systems. See Annex A for outline descriptions and characteristics of different levels of autonomy.

It is recognised that whilst the users of AMVs can be divided into three broad categories (research, commercial/industrial, and military), the vehicles themselves cannot be thus divided. In fact in recent years, more or less identical vehicles have been used by all three groups.

The *modus operandi* of these vehicles is also independent of their user, with large overlaps and interchange of ideas between the three communities.

d. AMV Classification

The technologies and uses of AMVs are evolving rapidly. At the time of writing, most commercially available AMVs are designed with the capacity for reconfiguration to suit a variety of different tasks, some yet to be conceived. Given the limited size of the market, vehicles have been designed to be attractive to as wide of a user base as possible. A classification by role is therefore not practical, whilst a classification by type would need to be constantly updated to encompass new variants.

The classification of vehicles is of most use to those organisations that are interested in quantifying and managing safety and risk. A classification is therefore proposed which uses vehicle mass together with its maximum velocity, i.e. its maximum possible kinetic energy. Therefore, the classification is related to the degree of harm which the vehicle may potentially cause to other users of the water space. These classifications, for risk management purposes, can be found in Annex B.

e. AMV Missions and Tasks

The type of mission which will be undertaken by an AMV is specified and assigned to one of the classes described in this section.

The main factor driving the requirement for this mission classification is how the role of the AMV will be perceived under the concept of the United Nations Convention on the Law of the Sea (UNCLOS). This classification is based around access to the different areas within the agreed maritime boundaries, such as the Exclusive Economic Zone (EEZ). The three different mission classes have significantly different access rights in the legally recognised range of maritime boundaries[5]. Even though some of the operational tasks may be identical in all three classes (e.g. oceanographic measurements), rights of access will vary depending on the status of the vehicle operator and the intended recipient of the acquired data. Examples of tasks that may be undertaken by an AMV are in Annex C, while examples of tools that may be used by an AMV are at Annex D.

(i) **Science/research for science missions (Class = S)** – Marine science AMVs have two main roles: as observation platforms that will, in certain scenarios, replace surface ships, drifting or moored buoys; and as a tool augmenting surface ships, towed vehicles, drifting or moored buoys, etc. While the first role is a major driver in marine science AMV development, operational use of AMVs independently of a surface vessel is likely to become commonplace as vehicles with longer endurance and greater autonomy are developed. It is expected that all data acquired during research missions will enter the public domain.

(ii) **Industrial/commercial missions (Class = C)** – Industrial AMV tasks, in general, embrace all those described Annex C, with the emphasis on seabed data acquisition. The main difference is linked to economic and fiscal considerations. The information or data collected via the industrial AMV is not, in general, freely available and is treated as commercially sensitive.

(iii) **Military missions (Class = M)** – Military AMV* tasks are, in general, akin to those described in industrial/commercial missions, however, there is a distinct and unique set of tasks that are only undertaken by military AMVs. These roles are in support of the military objectives.

(iv) **Regulatory authority (Class = R)** – This class covers AMVs when being operated by a government or other regulatory body in support of specific social objectives, e.g. search and rescue, identification, and any activity in support of the civil power.

---

* This CoP is not intended to apply to military AMVs carrying weaponry.

Code of Practice*

1. Legal Considerations

1.1. Compliance with Existing Framework
In the continued absence of law specific to AMVs, system manufacturers, owners, and operators will operate using the requirements described in volume two of this series[5], in conjunction with the explanatory notes contained in volume three[2], which is updated by Annex E.

1.2. Periodic Review
AMV system manufacturers, owners, and operators will undertake, through the auspices of the SUT, a biennial review of the law pertinent to AMVs and make revisions to this CoP as required. AMV system manufacturers, owners, and operators working within the guidelines set by this CoP will endeavour continually to refine its contents to reflect the current state of the art and industry best practice. Commitment to this CoP could be used as evidence of a responsible industry attitude in a court of law.

1.3. Precautionary Principle
AMV system manufacturers, owners, and operators should adopt a precautionary approach to the use of AMVs in all aspects. This can be paraphrased as 'if one is embarking on something new, one should think very carefully about whether it is safe (or legal) or not, and should not go ahead until reasonably convinced it is'[6].

---

* This Code of Practice is an evolving document, based on current best practice.

2. Contractual Considerations

2.1. Capabilities and Limitations
When planning a contract with a client, the AMV owner/operator should clearly describe the capabilities and limitations of both the system itself and the system under the influence of local conditions.

2.2. Delivery of Products
The format of the products to be delivered to the client should be clearly stated, together with the type of delivery media. Archive quality media should be employed. Unless specifically excluded, all raw data files should be delivered so that alternate methods of processing may be used. All data files should be time-stamped using a common time-reference source.

2.3. Fallback
Any contract for use of an AMV should clearly describe actions to be taken in the presence of uncontrollable conditions, e.g. bad weather, vessel damage. These may include alternate locations, missions, or methods of acquiring data.

3. Environmental Considerations
An assessment of the impact of the operation of the AMV on the environment in which it is to operate should be made. This should include the impact of loss or excursion of the vehicle, as well as planned operations. An appreciation of operating scenarios should be included (e.g. there are different risks entailed in working in a near-shore, high-traffic area compared with an offshore, deep area well away from traffic lanes; operations near and under ice bring additional risks).

There should also be third-party liability insurance in place, with operating procedures being agreed with the insurers.

3.1. Pollution
It is the responsibility of the vehicle manufacturers to minimise the use of materials in the construction of the vehicle and its subsystems which would present a hazard to the marine environment. Nonetheless, such materials will be used and should be encased or otherwise protected to prevent leaching in the event of vehicle loss.

A full list of such materials and their likely effects shall be available to the vehicle operator, including any materials present in subsystems added to the vehicle post-manufacture.

3.2. Vehicle Identification

All AMVs shall be clearly marked with:

a. The national flag of the owner, where it will be visible when the vehicle is surfaced; as large as is practical according to the size of the vehicle; and in multiple locations (in case of breakup following a collision) with the following information;
b. Full contact details of the vehicle owner; and
c. All hazards which may present themselves to a salvor.

The method of marking shall be able to withstand prolonged attack by chemical and biological agents.

3.3. Sonar, Lasers, and Other Electromagnetic Emitters

Most AMVs are fitted with an active sonar. Whilst the transmit power levels are generally low by design for conservation of vehicle power, levels can still be harmful to marine life, divers in the vicinity, or crew members if the sonar is activated on deck for test purposes.

The AMV operator shall verify that the transmit power, pulse duration, repetition rate, and frequency ranges of all emitters fitted to the vehicle fall within the recommended safe limits in accordance with the Environmental Impact Assessment (EIA) and published guidance by the relevant regulatory body or UNCLOS, for both marine life and human receptors.

A low-power mode for sonars is the best solution if on-deck testing is to be performed, as it also reduces risk of damage to the transducers from mismatched loading. The low-power mode should also be used if divers will be present within the AMV operating region.

3.4. Chemicals

In addition to the chemicals forming part of the vehicle's structure, including power pack and sensors, AMVs may carry chemicals as part of experiments.

Risk assessment should include the impact on the environment of both the planned and unplanned release of these. Substances planned for release should not be prescribed under national or international legislation, including Part 1 of the UK Environmental Protection Act 1990, EU Directive 76/464/EEC List II (Grey List) and EU Directive 76/464/EEC List I (Black List).

3.5. Mechanical Interaction with Third Parties

Risk assessment should cover the danger of moving parts to sea life, human beings in the water, and, upon surfacing, the effects of collision and damage to ships and other water users.

3.6. Environmental Impact Assessment

If permission is required to operate the AMV in a particular location, the authorising body will often require a formal EIA to be submitted with the application.

The assessment must be submitted in time to allow approval to be granted prior to contract acceptance for the project.

It is the intention of this document to assist both the regulatory bodies concerned and the operational unit in formalising the risk assessment.

4. Operational Considerations

Through the following paragraphs, the following terms are used as defined here:

Mobilisation – movement of the system from its home base to the operational area, whether a vessel or other launch site, where installation and test to operational readiness state will be carried out; if the system is a permanent vessel fit, then test to readiness only is required

Mission – deployment of an AMV from launch through recovery to offload of raw data

Task – acquisition of a particular data set according to client requirements; a mission may include several tasks, or a task may be spread over several missions

Post-processing – conversion of the raw acquired data to products specified by the client

Demobilisation – return of the system to a standby state after completion of a programme of work, either at its home base or on the vessel if permanent fit.

4.1. Planning

Planning for the use of an AMV can be divided into two sections, which can be viewed as static and dynamic.

'Static' planning will take place during contract meetings before the start of the programme of work. Issues that will be discussed and decided upon will be those that are not expected to be affected by change in the medium term.

'Dynamic' planning will take place at the beginning of the cruise and will be periodically reviewed during the cruise. It is expected that decisions will be amended in order to take account of changing conditions and to make the most efficient use of the AMV in fulfilling client expectations.

The following subsection list the minimum set of issues to be addressed.

4.1.1. Access to Operating Area

It is vital to obtain permission to operate within any particular area. Even in areas where formal permission is not required, ensure that Notice to Airmen (NOTAMS), Global Maritime Distress and Safety System (GMDSS), Automatic Identification System (AIS), Navigational Information Telex (NAVTEX), and local warnings (e.g. Channel 16) are all in operation. Furthermore, contact should be made with local users, the coastguard, and the military to advise of the nature and timing of operations.

The AMV operator must verify that formal permission has been granted.

It is recommended that permission is requested for an area larger than the planned area, to allow for unplanned excursions, navigational errors, or failure of vehicle.

4.1.2. Operating Site Suitability

Detailed drawings of the operating site and handling support facilities should be requested by the AMV operator at an early stage in project discussions. A site or vessel visit should be carried out, ensuring that key crew members or site personnel will be available to answer questions.

4.1.3. Installation Drawing

All equipment spaces, access spaces, lifting areas, transit areas, securing points and methods, and point loading data will be defined and marked up on an installation drawing. The drawing shall also specify electrical/

hydraulic/pneumatic power supplies required, as well as vessel data connections. This drawing will be submitted to the client, who will in turn seek approval from the site or vessel owner.

It shall be the responsibility of the client to liaise with the site or vessel owner to ensure that all facilities defined on the installation drawing are available for use at the time of mobilisation.

4.1.4. Custom Adapters and Fittings

It is likely that connection adapters and custom mechanical fittings will be required on a per site basis. The AMV operator shall ensure that these are available at the time of mobilisation.

4.1.5. Communications

Direct communications between the vessel bridge/dynamically positioned (DP) operators and operation and deck crew should exist throughout the mission, especially during launch and recovery.

Communications and the maintenance of communications during launch and recovery should take into account the following:

a. Line of sight between bridge crew and recovery crew

b. A backup system if radio communications are not possible or there is congestion on communal radio frequencies

c. The position of ship's machinery with respect to communications numbers, considering any automatically starting machinery and the recovery positions available

d. Liaison between surrounding vessels, managed by the bridge

e. A formal line of command between ship's boats, bridge, and the AMV crew.

4.2. Risk Assessment and Management

A critical element of the planning activity is risk assessment and the development of procedures to minimise the identified risks.

4.2.1. Forms of Risk Management

Three separate forms of risk management will be conducted, in addition to the environmental risk assessment, which may be required by authorising bodies:

Project Risk Management – to identify and manage issues such as cost, time scale, quality of delivered data and actions, and recovery in case of failure

Safety Risk Management – a statutory requirement on the operator that will identify all hazards and determine methods to negate or mitigate them. The application of different safety regimes will depend upon various factors, including the country of incorporation of the owning/operating company, the flag of the vessel from which the AMV is operated, and the territorial waters in which the operation is taking place. Operations within the North Sea or UK territorial waters will be subject to the Health and Safety at Work Act 1974 (HSWA), as implemented and enforced by the Health and Safety Commission (HSC) and the Health and Safety Executive (HSE). As part of HSWA, operators are required to complete a risk management review as part of their safety management system.

In instances where an AMV is operating as part of an offshore operation, an AMV operator maybe required to comply with additional safety legislation in the field. In UK territorial waters, additional legislation, as implemented by various EU Directives would include:

- The Management of Health and Safety at Work Regulations 1999;
- The Provision and Use of Work Equipment Regulations 1998;
- The Lifting Operations and Lifting Equipment Regulations 1998;
- The Personal Protective Equipment at Work Regulations 1992;
- The Manual Handling Operations Regulations 1992;
- The Health and Safety (Display Screen Equipment) Regulations 1992
- The Control of Substances Hazardous to Health (Amendment) Regulations 1999.

Environmental Risk Management – even if not a statutory requirement, an assessment of the risks to the environment of the operation that will be carried out (see Section 3).

4.2.2. Review and Update

It is to be expected that a particular AMV operator will be able to reuse most of the initial effort invested in the risk management systems, modifying and updating per the individual requirements of each client and/or cruise.

When a particular project is initiated, both of these risk management systems shall be used as live systems, in that they must be reviewed and updated on a regular basis appropriate to the phase of the project. The periodicity will be determined at the project planning stage.

4.2.3. Feedback and Effects

Outputs from the Safety Risk Management system will often impact the Project Risk Management system. This is normal and is permitted.

Outputs from the Project Risk Management system shall never be allowed to negatively impact the Safety Risk Management system.

Any adverse changes to the outputs of the Project Risk Management system shall be communicated to the client as soon as is practical.

4.2.4. AMV System Location and Integrity

An AMV may be located on a support ship, on a shore base, or at an underwater permanent or temporary base. Each location will have its own set of safety and procedural requirements for support equipment and overall integrity. For example:

Ship-based: the AMV and its support system may need to comply with national and international regulations pertaining to shipboard installations and operations.

Shore-based: the AMV will have to abide with the local health and safety regulations applicable to port/dock areas.

In-water-based: the integrity and reliability of the close-in navigation systems for pre-docking location and control will be crucial; the durability of the docking mechanism, the power and data transfer connections, and the procedures and mechanisms for undocking will also be important.

Any AMV operator will have to consider risk in the planning process. This will form the basis of a risk assessment that should be considered to be an essential formal document. The mission will be dependent upon the tools and the environment, together with operational equipment matters. The remainder of this document highlights these issues in order to assist in the preparation of the mission risk analysis.

AMVs will be subject to a wide range of operational considerations, and any particular AMV may be used in a number of applications with quite different risks and constraints.

The types of risk may include the risk of:

- Loss of life
- Injury
- Loss of AMV
- Damage to AMV
- Loss of other vessels and property
- Damage to other vessels and property
- Damage to environment
- Navigational hazard
- Theft of, or malicious damage to, the AMV
- Theft of data
- Political incident – e.g. incursion into third-party nation's Territorial Waters or EEZ.

The risk assessment should include: site- or mission-specific hazards; vehicle- and sensor-specific hazards; hazards arising from the deployment and retrieval method (ship-based, which may or may not require a swimmer; and shore-based, which may involve AMV navigation in confined waters); hazards to third parties; and the potential for environmental pollution. There may also be occasions when a formal permit may be required – e.g. under the provision of the UK Antarctic Act 1994 permission is needed from the Foreign and Commonwealth Office (FCO) for all expeditions south of 60ºS. Such permits may require a risk assessment to be carried out and a copy provided to the relevant authority.

All relevant safety practices should be complied with when fitting, testing, and operating the sensors and tools fitted to an AMV. The risk assessment for an AMV mission should include the risks associated with the sensors and tools, for example, high-power sonar systems (potential effect on both operators and marine mammals) or UV radiation (spectrophotometers and fluorometers).

Consideration should also be given to the risks associated with fitting sensors to the AMV; for example, some very delicate sensors may only be fitted by a swimmer once the vehicle is in the water. Particular risks to the AMV may arise from the pressure vessel design and specifications of a sensor or tool package. Careful planning and contingency analysis should be undertaken in such cases.

4.3. Influence of the Environment

The safe and efficient deployment and recovery of AMVs depends on suitable environmental conditions. This constraint will equally apply for both ship- and shore-launched AMV missions.

4.3.1. Operating Envelope

AMV planning will require an assessment to be made of the range of environmental conditions that will have significant impact on the safe deployment and recovery of the AMV, as well as the undertaking of its tasks. Because of the portability of AMVs, a single environmental operating envelope cannot be established.

The AMV operator will specify the operating envelope for the ship/AMV or shore-base/AMV combination for each specific operational environment.

4.3.2. Analysis and Forecasts

The use of both environmental analysis and forecasts prior to and during the period of each mission must be undertaken. This will allow the mission-specific environmental envelope to be considered and risks to the mission and the AMV operators to be determined.

The environmental analysis and forecasts will be used in conjunction with the Safety Risk Management and Project Risk Management systems in order to determine a course of action in adverse conditions.

4.3.3. Water Depth

All AMVs will have been designed or certified for a maximum working depth. Operators should ensure that a vehicle is never used below that limit. Operators should be aware of any maximum depth limits set in the AMV firmware, which may or may not relate to the maximum working depth.

When operating an AMV, consideration should be given to the following:

Minimum water depth: Can the vehicle mission be safely carried out in the depth of water present in the working area? When working in shallow waters, tide table and charts must be carefully analysed. Many vehicles cannot achieve maximum speed when at the surface and may not be able to navigate against a current.

Maximum water depth: If the water depth is greater than the maximum working depth of the vehicle, are the safety systems correctly set and are they working (for example, firmware limits on diving-depth altimeter functioning correctly)? Operators need to be aware that one or more subsystems on the vehicle may have shallower working depths than the vehicle itself, especially where they have been custom installed, and so the maximum working depth for a task may be set by one subsystem rather than the vehicle.

4.3.4. Temperature and Weather

Extreme air temperatures (high and low) may affect the reliability of electronics and cause material stress that leads to structural or mechanical damage. In Arctic conditions, water allowed to remain in crevices when the vehicle is on deck will freeze, possibly causing damage to joints and seals. In low latitudes, solar radiation can heat unprotected plastic materials to their softening point and result in high temperatures within sealed containers. Hydraulic fluids and lubricants should have stable properties over the intended temperature range.

Weather is likely to affect any surface support craft more than it will affect the AMV during its mission. Sudden deterioration of weather may force the surface craft to leave the area of operations, thus the AMV mission plan and programming must include procedures to deal with such a situation.

Adverse weather conditions during launch and recovery operations are particularly hazardous, and careful monitoring will be required to ensure the safety of personnel and equipment. If operating in cold conditions, care should be taken to minimise problems arising from icing of control surfaces, sampling tubes, or actuators. Particular care should be taken when recovering an AMV in polar waters when any relatively warm seawater trapped in the vehicle may quickly freeze upon exposure to the colder air. During hot weather, an AMV on deck may heat up significantly, and its initial in-water buoyancy may be affected.

4.3.5. Water Density (Salinity)

Water density variations affect the buoyancy of an AMV. Good practice will include sensitivity analyses of overall vehicle buoyancy and trim over the range of density (salinity/temperature/pressure) conditions likely to be encountered. Particular attention should be given to:

a. AMV missions departing coastal locations (e.g. fjords, intra-coastal waterways, or estuaries where fresh, low-density surface waters may be found)

b. Polar waters, where low-density waters may occur due to ice melting, or higher density waters may be found when brine rejection occurs during ice formation

It is probable that the overall compressibility of the AMV will not match exactly that of seawater. The AMV will therefore gain or lose buoyancy with operating depth. Operators should be aware of the impact that such a buoyancy change might have on operations (e.g. changes in mean pitch angle, energy consumption).

4.3.6. Pollutants

The presence of man-made or natural petroleum products can cloud optical lenses and damage plastic materials. Gas can affect visibility, block sound transmission, and cause sudden loss of buoyancy. If pollutants are present, precautions should be taken to protect the in-water portions of vehicles and any personnel who handle the AMV during launch, recovery, and maintenance.

4.3.7. Currents

Currents can cause considerable problems in AMV operation, not only affecting the trajectory of the vehicle, but also diluting the accuracy of acquired data. Wherever possible, historical data, simulations, or analyses should be obtained for the working area, whilst recognising that currents vary with location, depth, meteorological conditions, and time. Mission planning must take account of the range of currents likely to be encountered, and the mission track should be designed to be achievable under the prevailing conditions. Any mission programme should be structured such that failure of the vehicle to reach a given way-point due to adverse current will result in minimal impact to the overall mission plan.

When planning a mission, careful consideration should be given to the effect of current on the vehicle when it is on the surface (e.g. when waiting for a navigation fix). The time taken for a fix may be unpredictable, and the mission plan should account for drift during such a fix. This effect of current is likely to be of greater importance in coastal areas and estuaries.

Vertical currents associated with internal waves, internal tides, and soliton packets will influence AMV motion and should be considered when working in areas likely to be affected – for example, near the shelf break or in straits.

All of these issues regarding operating AMVs in currents will be even more critical if the AMV is required to dock into a fixed or moored docking station, or if two or more AMVs are required to work cooperatively.

4.3.8. Sea State and Swell

Sea state can affect every stage of an AMV operation.

Safety should always be considered carefully when launching or recovering an AMV, particularly from a support vessel in rough seas. AMV operators should understand the effect of a heaving support ship on a cable attached to a relatively motionless AMV and be aware that the AMV handling system can be overloaded, or that personnel on deck may be exposed to a risk of accident.

In rough conditions, personnel involved with launch and recovery must wear all necessary personal protective equipment and fully understand their own role as well as the roles of others involved in the operation. Good communication is vital to avoiding accidents.

The master has a well-established legal responsibility for the safety of the crew, all other embarked personnel, and the ship. During hazardous operations, which will include launching and recovery of the AMV, the AMV supervisor will decide whether conditions are suitable and the equipment is in proper operational readiness, but the master will be ultimately responsible for the final decision to deploy or recover the AMV.

In certain situations, purpose-built deployment systems, utilising motion compensation systems, can reduce the effect of wave action, thereby enabling AMV operations to be conducted in higher than normal sea states while maintaining safety standards.

Consideration should be given to the selection of the minimum depth command for the AMV when operating in high sea states to avoid unintentional surfacing.

4.3.9. Seabed Characteristics

When planning AMV operations, enquiries should be made about local seabed conditions and topography. On occasion, a reconnaissance mission at a safe height might be appropriate. Rocky outcrops, wrecks, or seabed equipment may make collisions more likely. Operating the vehicle in a terrain-following mode, either on continuous terrain-following or when programmed to run to a set distance off the bottom, will increase the risk of collision. The safe maximum terrain-following height or turning height off the bottom should be informed by knowledge of the seabed characteristics.

4.4. Operating Procedure

Operating, safety, emergency, and maintenance procedures should be put in writing and agreed by interested parties (e.g. operator, manufacturer, insurer, and owners of seabed installations, as well as any hirers).

The AMV system should be provided with an operations handbook that should be updated on a regular basis to reflect practical experience with the AMV. The handbook should be easily available to the AUV operating team.

The team operating the AMV should have a designated leader and, if operating continuously, shift leaders. Each team member should be aware of his or her individual responsibility and any backup roles in case of illness or emergency. There should be clear demarcation of responsibilities for all stages of the operational cycle of an AMV, with individual responsibilities identified and individuals to be guided by the operating procedures as discussed previously. The mission documentation should also include:

a. Scope of work, timetable, operating area, permissions, and clearances, reporting requirements
b. Mobilisation and demobilisation plan, any customs documentation
c. Designated personnel and duties
d. Emergency contact details – personal and for vehicle relocation, e.g. through Advanced Research and Global Observation Satellite (ARGOS)
e. Contingency analysis and plans
f. Scientific data reporting responsibilities – e.g. Intergovernmental Oceanographic Commission (IOC) Cruise Summary Report (CSR) forms
g. Risk assessments (see 4.2).

4.5. Handling Systems
4.5.1. Introduction
Experience shows that the launch and recovery phases of an AMV mission pose particular hazards to the vehicle, any support platform or vessel, and personnel.

4.5.2. Documented Procedures
Due care and attention must be given to the design, construction, and use of AMV handling systems. The vehicle handbook should include explicit

step-by-step instructions for deployment and recovery of the vehicle under all anticipated conditions of use. These instructions should include procedures for recovering the AMV when its onboard systems have failed or flooded. If, for any reasons, the documented procedures cannot be used, a written – and preferably a video – record of the 'non-standard' procedure should be obtained.

4.5.3. Sea State
Particular attention should be given to the surge loads that may be encountered during launch and recovery arising from the different dynamic characteristics of the AMV and the support ship, particularly in heavy seas. Entrained water or a flooded vehicle will pose particular problems.

If possible, the vessel shall be positioned such that the launch/recovery takes place on the lee side.

4.5.4. Supervision and Communication
The person in charge during launch and recovery should be clearly identified, should be one of the AMV support team, and should be in constant communication with the watch-keeping officer in the wheelhouse. The AMV team leader, in conjunction with the vessel's master, must ensure that the designated launch/recovery leader and the watch-keeping officer are fully aware of, and conversant with, all the practices and procedures laid down in the vehicle handbook and other relevant documents.

4.5.5. Maintenance
Although not specifically produced for AMVs, guidance on handling systems examination, testing and certification is available in Reference 4.

4.6. Pre- and Post-Mission Procedures and Checks
The AMV operator should develop a checklist for pre- and post-mission procedures and checks. This checklist should be used on each mission and a copy included in the mission log. The team leader should ensure that team members either complete all operations included on the checklist, or initial why a particular operation was not necessary (for example, a listed subsystem may not be installed). The checklist should at least cover:

    a. The testing of all AMV subsystems or modules, which should be carried out before the vehicle enters the water; these tests should, as far as is practicable, extend to the installed sensors and instruments

b. The checking of key mechanical components and control surfaces, which should be carried out before deployment, and a visual check of sternplane and rudder 'zero' positions, which should be made, if possible, after deployment but before the vehicle's first missions

c. The comparison of the AMV position from its own navigation device(s) with shipboard or other reference before the vehicle is allowed to start its mission

d. The testing of navigation warning beacons, such as flashing lights, radio DF transmitters, and satellite transmitters, and the checking of the condition of all antennas and housings.

Allied to this, there should be procedures to cover vehicle programming and system checks. Most AMV operational incidents occur as a result of operator (i.e. human) error. A rigorous independent checking procedure is therefore necessary to minimise/eliminate such errors (e.g. vehicle programming checked independently with no input from the primary programmer).

4.7. Communications and Relocation

4.7.1. Team Briefing

All personnel (AMV team and all ship's watch-keeping officers [deck and engine], including the master where applicable) involved in the AMV operations should be fully aware of the plans and operating procedures. The AMV team leader should discuss contingencies and any unusual situations that may arise with all involved, especially for deployment and recovery.

4.7.2. Radio Communications

Adequate means shall be available to communicate between team members and vessel crew, with authorities (e.g. coastguard), other vessels, and any emergency relocation service being used (e.g. ARGOS). The team leader must have access at all times to such methods of communication.

The AMV team leader shall verify that all licences, subscriptions, and personnel qualifications are current for the radio communications systems to be used.

4.7.3. Vehicle Relocation

It is advisable to incorporate an acoustic beacon in an AMV to function as an emergency location aid and also as a hazard warning. At its simplest, the beacon could have a continuously running pinger, although there are benefits in using a transponder, such as the ease of obtaining a range to the vehicle. The beacon should use a transmission frequency in accordance with industry norms, such as within the 27–33kHz or 10–12kHz band. The characteristics of the beacon should be notified to relevant authorities e.g. Hydrographic Office, Maritime and Coastguard Agency.

If radio-direction finder beacons are installed on the AMV, then suitable reception equipment should be available, installed, and tested on the support vessel or shore base.

4.7.4. Neighbouring Activities

If AMV operations are being interleaved with other activities onboard a vessel (e.g. lowering instrumentation on wires, towing equipment), independent communication channels shall be established between the AMV team leader and the leader of such other activity to the vessel bridge, to ensure that safety is not compromised. Contingency plans should be available for both activities in the event of unforeseen situations with the other activity.

4.7.5. Communications with the AMV

The AMV may incorporate one or more communications links to a support vessel or a shore base. The communications link may use acoustics or radio, in which case it may be terrestrial or via a satellite. A communications link may be used for data transmission from the AMV, but it may also be used to download data and mission profiles to the AMV. The AMV operator should be aware of, and should guard against, any security risks associated with the communications link. The possibility of unauthorised persons obtaining control of the AMV over the communications link should be very carefully considered. Procedures such as password protection should be considered for mission updates. If used, passwords should not be sent over the link in plain text, but should be encrypted.

4.8. Navigation

If the AMV is to transit on the surface in restricted waters, then procedures must be adopted to ensure adequate navigation.

The AMV operator should incorporate, within the vehicle's operating procedures, an analysis of the positional accuracy when submerged sufficiently to demonstrate that the vehicle is not a hazard to other vessels or subsea installations.

If multiple AMVs are used, or if AMVs are used in conjunction with ROVs or other devices liable to cause interference to a position location system, then the risk should be analysed and documented, and a mitigation plan developed. The operating procedures for the AMV should state how the hazards identified in the mission risk analysis will be controlled or mitigated to an acceptable low level. Consideration should be given to the AMV carrying suitable active and passive acoustic and/or optical sensors to provide collision avoidance suitable for the missions to be undertaken. These may include a simple altimeter (to avoid collision with the seabed), a multiple-beam sonar for forward-looking collision avoidance, an acoustic pinger to aid emergency relocation and provide a proof of presence, or a transponder system for tracking from a support vessel. Section 5.7 also deals with navigation systems and beacons.

4.9. Manuals and Documentation

A major factor in the safe and efficient operation of AMVs is the provision of a comprehensive set of up-to-date written material. It is the team leader's responsibility to ensure that each AMV system is supplied with the necessary documentation, including:

- Operational manual
- Safety management system
- Technical manuals for system equipment
- AMV diary/report book/mission log
- Planned maintenance system
- Repair and maintenance record/spare parts inventory
- Pre-/post-mission checklist
- Recovery/launch environmental envelope
- Operational environmental envelope
- Emergency contact details (including insurance and water space management).

4.10. Hazards to Operating Personnel

4.10.1. Safe Use of Electricity

AMVs may use electrical power sources with sufficiently high voltages and current capacity to constitute a shock hazard. Hazardous voltages may exist within the main power pack, the propulsion motor and its associated circuitry, and within some sensor payloads, such as sonars. The hazard presented by such equipment when used ashore may be modified when used at sea, for example, due to the pervasive presence of salt water.

Batteries housed in sealed pressure vessels constitute a hazard. Design and operating procedures should be developed to minimise the hazard from gas build-up and possibly high temperatures. Novel energy sources such as fuel cells, with reactants under pressure or in cryogenic storage, or high-temperature batteries (such as sodium-sulphur) require particular care. Risk assessments should examine all issues related to the safe use of electricity on the AMV. Guidance may be obtained from documents such as *Safety procedures for working on high voltage equipment*[7].

4.10.2. Moving Parts

The moving parts (including rotating machinery, drop weights, manipulators, planes, and folding components) of an AMV constitute a hazard ashore, on a ship, and when in the water (see Section 4.10.4). The propeller may start without warning; there may or may not be a propeller guard in place; the propeller may start upon a radio, umbilical, or acoustic command; or it may start due to faulty hardware or software. Recognised hazard warnings for rotating machinery should be permanently fixed to the vehicle, and a safe operating procedure should be developed and documented (see Section 4.2).

Other moving surfaces, such as sternplanes and rudders, may also constitute a hazard, as they may operate without warning.

4.10.3. Lasers

To penetrate water, high-power lasers are often used. There is a serious risk to personnel on deck if these are used in air and possibly to divers if they get too close to them in water. The strong attenuation of light, especially in the infra red, greatly reduces the risk in water. The correct safety precautions must be taken if a class III laser or above is to be used. Approved warning labels should be attached to the hull.

4.10.4. Use of Divers

There may be occasions when the use of divers during launch or recovery may be required. If so, a risk assessment should be completed as part of the mission documentation (see section 4.2 and 4.9), bearing in mind, amongst other factors, the sea state; the prevailing weather and water temperature; training and experience of the diver(s); the nature of the intervention (routine or emergency) and support craft capability. Divers should be aware of the potential for machinery to start up without warning (see section 4.10.2).

4.10.5. Manual Handling

AMVs pose similar manual handling hazards to other large items of equipment. The confined space within an AMV may present additional hazards when mounting or dismounting large or heavy modules.

5. Equipment Certification and Maintenance

Various standards and codes are used to examine, test, and certify offshore plant and equipment, and the requirements of those who are competent to carry out such examinations, tests, and certifications have been established[5,8–11]. Much of the equipment used in an AMV operation should comply with at least these standards. Relevant certificates (or copies) should be available for checking at the work site.

AMV equipment is used under extreme conditions, including frequent immersion in salt water. It therefore requires regular inspection, maintenance, and testing to ensure it is fit for use, i.e. that it is not damaged or suffering from deterioration. Regular maintenance is an important factor in ensuring the safe operation of an AMV system.

The frequency and extent of inspection and testing for all items of equipment used in an AMV operation, together with the levels of competence required of those carrying out the work, should be identified by the AMV users.

Procedures for vehicle repair/upgrade/testing should be agreed in writing between manufacturer and operator. Important examples include:

a. Maintenance procedures and intervals agreed between operator and manufacturer

b. If a vehicle pressure hull is split and remade, how long the vacuum/positive pressure should be monitored before vehicle deployment, so as to check the seal

c. Software function testing for software upgrades.

5.1. Planned and Periodic Maintenance

It is important that the AMV user establishes a system of planned maintenance for plant and equipment. Such a system may be based on the passage of time, amount of use, manufacturers' recommendations, or previous operational experience. Ideally, it will be based on a combination of all of these.

For each major unit, the planned maintenance system should identify the required frequency for each equipment item and who should do the work. The individual involved will then need to complete a record of the work, either on paper or computer, and then 'sign it off'.

5.1.1. Design, Testing, and Periodic Examination of Handling Systems
AMV handling systems have been the cause of injuries to AMV personnel. Due care and attention is vital at all times during their operation.

A list of safe AMV operating parameters for the handling system should be readily available. Operators should also be aware of the loads to which the system is subjected during normal operations. Standard welding procedures and non-destructive testing (e.g. dye penetrant) should be applied to all load-bearing fastenings associated with tie down, both before and after load testing.

Although AMV handling systems are not specifically covered by classification society rules, the Association of Offshore Diving Contractors (AODC, now the IMCA) guidance note[8] provides information that helps to avoid confusion over design, testing, and periodic examination standards. The note extrapolates relevant UK regulations to handling systems, but may be used in countries where local or national regulations do not provide alternative requirements.

All lifting equipment and control systems should be examined by a competent person, and the test certificate checked before the equipment is used for the first time, after installation at another site, and/or after major alteration or repair which may affect its integrity.

Regular examination every six months is also recommended. Any additional testing specified should be at the discretion of the competent person.

AMV lifting cables should be provided with test certificates confirming the safe working load (SWL). Details of the required SWL for lifting cables are given by the IMCA[9]. The SWL should not be exceeded during operations and should include the AMV and any components that hang from the lifting cable (including cable weight in air). The condition and integrity of any lifting cables should be checked at six-monthly intervals, or more frequently if circumstances require it.

A dedicated lift cable, or an umbilical which is used for lifting an AMV, should be re-terminated at intervals of twelve months, and examined and re-tested by a competent person. Detailed guidance is given in IMCA R 003[11].

The lifting and lowering winch system should be rated by the manufacturer for an SWL based on the dynamic load of the AMV in air, plus any additional components and trapped water. The rated sea state or heave for the system should also be clearly marked in the operating manual. The test certificate for the overload test of the winch's lifting and braking capacity should be checked.

All lifting gear, such as sheaves, rings, shackles, pins, and eyes, should have test certificates when supplied and be examined at six-monthly intervals thereafter. The certificates should show the SWL and the results of load tests undertaken on the components to two times the SWL.

Many complex action sequences are required during an AMV operation, and there is a risk that steps may be omitted or performed out of sequence. A suitable way to ensure the thoroughness of such sequences on each occasion is to use a checklist that requires relevant personnel to demonstrate correct completion by ticking a box.

AMV users should prepare and authorise the use of such checklists as part of the planning for operations.

5.2. Spare Parts

AMV operations are often undertaken in remote areas. AMV users should therefore ensure that an adequate supply of spare serviceable parts is available, particularly for those items that are essential to continued operation, safety, and recovery.

5.3. Equipment Register

An equipment register should be maintained at the work site, together with copies of all relevant certificates of examination and test. It should contain any relevant additional information, such as details of design limitations (e.g. maximum weather conditions).

6. Personnel

The AMV should only be operated by approved personnel (e.g. those who have attended appropriate training courses or have the relevant experience).

6.1. Qualifications and Experience

The following minimum qualifications are regarded as the normal industry-accepted standards for AMV personnel:

a. Manufacturers' operators and/or supervisors course
b. A responsible attitude to work, as AMV personnel are required to work in a team and may be subjected to the pressures of long shifts
c. A verified logbook recording work experience, training, etc., especially if gained on AMVs
d. HSE and other appropriate statutory training certificates for working in the relevant marine environment
e. A medical certificate for working in the relevant marine environment
f. Appropriate navigational qualifications (not required for all team members) that are sufficient for the mission plan to be verified for navigational integrity.

Possession of a certificate does not in itself demonstrate competence for a specific operation. When assessing competence, AMV users should establish the overall experience of the candidate in operations of similar complexity. The team leader should also take into account the equipment and facilities that will be available, any potential risks, the availability of support in an emergency, and any other relevant factors. The candidate's logbook should be used to assess competence.

Extensive experience may substitute for academic qualifications, but each case should be treated individually.

The medical fitness of personnel should meet any relevant company, local, or national requirements for offshore topsides personnel. This may well be related to their age, but should require a medical examination every three years as a minimum.

6.2. Team Size

Safety of personnel is paramount during operations and maintenance, and it is the responsibility of the team leader to provide a well-balanced, competent team of sufficient size to ensure safety at all times. When selecting the team size, the AMV team leader should consider:

a. Nature of the work being undertaken
b. Deployment method
c. Number of vehicles deployed
d. Location

e. Vehicle classification
f. Operational period (12 or 24 hours per day)
g. Handling of any foreseeable emergency situations.

The AMV operator should provide a sufficient number of competent and qualified personnel to operate all the equipment and provide support functions to the AMV team, rather than relying on personnel provided by others for assistance (e.g. clients, ship crews, etc.). For safe operation, the team may also need to include additional deck support, as well as other management or technical support personnel, such as project engineers or maintenance technicians.

Personnel not employed by the AMV operator can create a hazard to themselves and others if they lack familiarity with established AMV users' procedures, rules, and equipment. Their competence and suitability should be carefully considered before their inclusion in the AMV team.

There may be exceptions to this, for example an AMV installed on long-term contract on a support vessel manned by suitable technicians employed by the vessel owner. In such circumstances, these personnel, whose principal duties may be associated with diving or the ship, could form part of the AMV team. Such an arrangement should be confirmed in writing, together with the responsibilities of these individuals.

The variance in vehicle type and tasks, together with advances in technology, make it difficult to offer anything more than issues for consideration in this CoP. Furthermore, it is not the aim of this document to remove the responsibility for safe operations from the team leader and his or her employer. Actual team sizes should be decided after the completion of a risk assessment.

Safe working practice dictates that personnel should not work alone when dealing with:

a. High voltage equipment – guidance in AODC document[7]
b. Heavy lifts
c. High-pressure machinery
d. Umbilical testing – where relevant (e.g. AMV/ROV hybrid)
e. Potential fire hazards – welding, burning, etc.
f. Epoxy fumes, etc.

Individuals in an AMV team may carry out more than one duty, provided they are qualified and competent to do so and that their different duties do not interfere with each other. Overlapping functions should be clearly identified in operational procedures.

Trainees may form part of the team, but should not normally be allowed to take over the functions of the person training them, unless that person remains in control and is present to oversee their actions, and the handover does not affect the safety of the operation.

6.3. Working Periods

Work should be planned such that each person is normally expected to work no more than 12 hours in any 24-hour period. The team leader will be responsible for ensuring that the roster provides for safe working and adequate meal and refreshment breaks. The roster should provide for a period of at least 8 hours of unbroken rest in each 24-hour period. See also Standards of Training, Certification and Watchkeeping (STCW)*.

6.4. Training

AMV users should ensure that their personnel have received necessary safety and technical training in line with any relevant legislation and, where appropriate, have met specific contractual conditions or requirements.

6.4.1. Safety Training

Safety training should include:

  a. Courses on survival, first aid, and firefighting
  b. Installation or vessel-specific safety induction covering possible hazards at work and while responding to emergencies
  c. Task-specific safety outlining the hazards associated with tasks such as working overside
  d. Refresher training at regular intervals.

6.4.2. Technical Training

AMV personnel should attend technical training courses, as appropriate, in order to gain a sound knowledge of the operation and maintenance of AMVs and associated equipment. Details of courses attended should be recorded in each individual's personal logbook and in the employer's personnel records.

---

* See www.stcw.org

6.5. Language

Personnel tend to revert to their own language in emergencies. If team members do not have the same mother tongue, this can be hazardous. The AMV plan should state which language is to be used during the project, and all team members should be able to speak fluently and clearly to each other at all times, particularly during emergencies.

For operations that are to take place in inshore waters, the team or vessel crew should include one member who is able to communicate in the dominant local language.

6.6. Logbooks

The AMV team leader should maintain an operational logbook in a standard form that documents all activities connected with the AMV missions, including, but not limited to:

a. Times of significant events
b. Record of all test certificates relating to the AMV and its subsystems, e.g. pressure vessels, lifting tackle, handling system
c. Originals of all checklists used in preparing the vehicle for its missions
d. Copies of the mission script used for mission planning and programming, signed and dated by the mission programmer and the team leader
e. Copies of any permits issued in connection with the missions
f. Copies of track plots from vehicle tracking or navigation systems, signed and dated
g. A full record of any emergencies or unforeseen problems encountered during the missions, with an analysis of the impacts or solutions adopted
h. An account of the energy budget for the vehicle, to ensure an adequate margin of safety for subsequent missions
i. Environmental conditions.

6.7. Allocation of Responsibilities

While attached to the ship, the vehicle is the legal responsibility of the master (as a piece of ship's equipment). When operating autonomously, the vehicle is expected to be the responsibility of the team leader. The hand-over from one to the other should be recognised explicitly and recorded as such in both the AMV and the deck logs.

6.7.1. Team Leader

The team leader is responsible for the operation that he or she has been appointed to supervise and should only hand over control to another suitably qualified person. Such a hand-over should be entered in the relevant operations' logbook.

The team leader with overall responsibility for the operation is the only person who can order the start of an AMV operation, subject to appropriate work permits, etc. Other relevant parties, such as the ship's master or the installation manager, can, however, advise the AMV supervisor to terminate work for safety or operational reasons.

To ensure that the AMV operation is carried out safely, the team leader should adhere to the following points (for convenience 'he' and 'his' are used as abbreviations for he/she and his/her):

a. He should be satisfied that he is competent to carry out the work, and that he understands his area and level of responsibility and who is responsible for any other relevant areas. Such responsibilities should be included in relevant documentation.

b. He should be satisfied that the personnel he is to supervise are competent to carry out the work required of them.

c. He should check that the equipment he proposes to use for any particular operation is adequate, safe, and properly certified and maintained. He can do this by confirming that the equipment meets the requirements of this CoP.

d. He should ensure that the equipment is adequately checked by himself or another competent person prior to use. These checks should be documented, for example, on an operation checklist and recorded in the operations' log.

e. When the operation uses, or plans to use, complex or potentially hazardous equipment, he should ensure that the possible hazards are evaluated and fully understood by all parties and that training is given if required. This will be carried out as part of the risk assessment during the planning of the operation and should be documented. If the situation changes, further risk assessment should be considered. The team leader will meet his responsibilities by ensuring that this documentation exists and by following any guidance contained in it, for example manufacturer's instructions.

f. He should ensure, as far as is reasonably practicable, that the operation he is being asked to supervise complies with the requirements of this CoP. Detailed advice on how he can ensure this is given in various sections of this code.

g. He should establish that all relevant parties are aware that an AMV operation is going to start or continue. They will also need to obtain any necessary permission before starting or continuing the operation, normally via a permit-to-work system.

h. He should have clear audible and, if possible, visual communications with any personnel under his supervision. For example, a team leader will be able to control the raising and lowering of an AMV adequately if there is a direct audio link with the winch operator, even though the winch may be physically located where the supervisor cannot see it.

6.7.2. Team Members

The team size will be in accordance with Section 6.2. Specific roles should be according to the risk assessment and role and task on which the team is employed. Specific roles should be defined and allocated to the team when and where appropriate. A deputy to the team leader should be identified.

6.7.3. Other Personnel

The actions of other personnel can have a bearing on the safety of the AMV operation even though they are not members of the team.

a. These other personnel may include:

(i) The client who has placed a contract with an AMV company for an operation; the client will usually be the operator or owner of a proposed or existing installation or pipeline, or a contractor acting on behalf of the operator or owner; if the operator or owner appoints an onsite representative, then this person should have the necessary experience and knowledge to be competent for this task

(ii) The main contractor carrying out work for the client and overseeing the work of the AMV operating company (the operator), according to the contract

(iii) The installation manager who is responsible for the zone inside which AMV work is to take place

(iv) The master of a vessel, or floating or fixed structure from which

AMV work is to take place; the master controls the vessel and has overall responsibility for its safety and all personnel.

b. These personnel should consider the actions required of them as follows:

(i) They must provide facilities and extend all reasonable support to the AMV team leader in the event of an emergency.

(ii) They should consider whether any underwater or above-water items of plant or equipment under their control may cause a hazard to the AMV team. Such items include water intakes or discharge points causing suction or turbulence, gas flare mechanisms that may activate without warning, or equipment capable of operating automatically. The AMV team leader should be informed of the location and exact operational details of such items in writing and in sufficient time to account for them in the risk assessments.

(iii) They should ensure that other activities in the vicinity do not affect the safety of the AMV operation. They may, for example, need to arrange for the suspension of supply boat unloading, overhead scaffolding work, etc.

(iv) They should ensure that a formal control system, for example a permit-to-work system, exists between the AMV team, the installation manager, and/or the vessel master.

(v) They should provide the AMV user with details of any possible substance likely to be encountered by the AMV, and therefore the AMV team, that would be a hazard to health, e.g. drill cuttings on the seabed.

(vi) They will also need to provide relevant risk assessments for these substances. This information should be provided in writing and in sufficient time to allow the AMV user to carry out the relevant risk assessments.

(vii) They should keep the AMV team leader informed of any changes that may affect the AMV operation, e.g. vessel movements, etc.

c. When operating from a DP vessel, the DP operator must inform the AMV supervisor of any possible change in position-keeping ability as soon as it is known. A duplicate set of DP alarms and clear instructions as to their meaning in the AMV control centre would be of value.

References

1. International Marine Contractors Association (IMCA). (1997). *Code of practice for the safe and efficient operation of remotely operated vehicles*, IMCA R004, Rev. 2. London: IMCA.

2. Brown, E. D., and Gaskell, N. J. J. (2000). *The Operation of Autonomous Underwater Vehicles series. Volume Three: The Law Governing AUV Operations, Questions and Answers.* London: Society for Underwater Technology, 82pp.

3. Griffiths, G., Mcphail, S. D., Rogers, R. J. and Meldrum, D. T. (1998). Leaving and returning to harbour with an autonomous underwater vehicle. In: Proceedings, Oceanology International '98, Brighton, Vol. 3, 75–87.

4. Rogers, R. J. (1998). Autonomous Underwater Vehicles (AUVs) – Some policy issues. Proceedings, UUVS '98, September, Southampton, UK.

5. Brown, E. D., and Gaskell, N. J. J. (2000). *The Operation of Autonomous Underwater Vehicles series. Volume Two: Report on the Law.* London: Society for Underwater Technology, 274pp.

6. Saunders, P. (2000). Use and abuse of the precautionary principle. London: Institute of Science in Society. Available at http://www.ratical.org/co-globalize/MaeWanHo/PrecautionP.html (accessed on 29 May 2009).

7. Association of Offshore Diving Contractors (AODC). (1993). *Safety procedures for working on high voltage equipment*, AODC 060. London: IMCA.

8. Association of Offshore Diving Contractors (AODC). (1994). *The initial testing and periodic examination, testing and certification of ROV handling systems*, AODC 036 Rev. 1. London: IMCA.

9. International Marine Contractors Association (IMCA). (1997). *Guidance on termination of load bearing umbilicals or lift cables use in ROV handling systems*, IMCA R003. London: IMCA.

10. Association of Offshore Diving Contractors (AODC). (1985). *Code of practice for the safe use of electricity under water*, AODC 035. London: IMCA.

11. International Marine Contractors Association (IMCA). (1996). *Basic level of competence to be met by ROV personnel*, IMCA R002. London: IMCA.

Glossary of Terms

AIS	automatic identification system
AMV	autonomous marine vehicle
AUV	autonomous underwater vehicle
AUVLWG	Autonomous Underwater Vehicle Legal Working Group of the SUT
CoP	Code of Practice
EEZ	Exclusive Economic Zone
GMDSS	Global Maritime Distress Safety System
IMCA	International Marine Contractors Association
IMO	International Maritime Organisation
NAVTEX	An international, automated system for distributing maritime navigational warnings, weather forecasts and warnings, search and rescue notices and similar information to ships.
NOTAMS	Notice to Airmen
ROV	remotely operated vehicle
SWL	safe working load
Team leader	The individual legally and managerially responsible for the actions of the AUV operating team and who must be on site at all times, perhaps using a nominated deputy.
UNCLOS	United Nations Conventions on the Law of the Sea
UUV	unmanned underwater vehicle (includes AMVs and ROVs)

Annex A: Degrees of Autonomy*

Levels of autonomy are defined as:

None

A vehicle with no autonomy will not have any form of automatic control. All of its functions will be directly controlled by a mechanical, electrical, optical, electromagnetic, or acoustic control link.

Some directly operated vehicles will, however, have a simple 'emergency abort system'. This is a device fitted to the vehicle that will react to the failure of a primary vehicle system. The device will initiate a sequence that will abort the vehicle mission, putting the vehicle into an emergency recovery mode that will facilitate safe recovery of the vehicle. Primary vehicle systems are considered to be:

- The vehicle power supply
- Navigation system
- Control system.

Basic

A vehicle with a 'basic' level of autonomy will have the ability to navigate a pre-programmed course and maintain a planned depth or altitude using a

---

* Please note: this CoP applies to *basic* and above only.

self-contained control system. The vehicle will include an emergency abort system that will be controlled by a vehicle health monitoring system, which will continuously monitor the critical subsystems of the vehicle to ensure that each system is operating correctly. On detection of a recoverable fault, the health monitoring system is to instigate the repair; if the fault is determined not to be recoverable, the system will instigate the abort sequence.

Intermediate

A vehicle with an 'intermediate' level of autonomy will have the characteristics of the basic level, plus an obstacle avoidance system that will change the vehicle trajectory to avoid objects that could damage the vehicle. A vehicle with this level of autonomy may also have control of sensors and their deployment, as well as intelligent interpretation of data gathered.

Advanced

A vehicle with an 'advance' level of autonomy will have the characteristics of the intermediate level, plus the ability to read the data gathered from the on-board sensors and adjust the pre-programmed mission plan to optimise the mission results. It is expected that new trajectories planned will be subject to maximum operational boundaries. These boundaries will be programmed as part of the vehicle mission.

Highly Advanced

A vehicle with a 'highly advanced' level of autonomy is one which is considered to have a level of independent control that is beyond all other categories. Examples of this are vehicles that have:

- Co-operative behaviour between multiple vehicles
- Task-based mission planning, where the AUV would determine its course or its sequence of operations for itself. It may trade efficiency for risk, and it may optimise its work independently of higher command. This implies that the operator/owner may not have prior or current knowledge of where the vehicle would be at any time.

ANNEX B: CLASSIFICATION FOR RISK MANAGEMENT PURPOSES

The recommended divisions are by size. Based on the existing population of AMVs, the range is divided into four arbitrary groups: I, II, III, and IV. The divisions are made on the basis of the maximum kinetic energy that the vehicle may possess, and calculated from the simple scaled formula of $m.v2$, where m (kg) is the total enclosed volume including flooded sections (i.e. the moving mass assuming neutral buoyancy *or* the displacement in the case of a non submersible vehicle), and v (m.s-1) is the maximum velocity of which the vehicle is capable in the horizontal plane.

The proposed divisions are:

- I = 0–100 — this grouping will be mostly gliders
- II = 101–1000 — this grouping will be small AMVs (e.g. Gavia, Remus 100)
- III = 1001–10 000 — this grouping will be large AMVs (e.g. Hugin, Autosub)
- IV = 10 001 and above — this grouping will be mostly fast surface vehicles, but can also include AUVs (e.g. Urashima, Marlin)

ANNEX C: EXAMPLE TASKS

Some of the tasks which can be performed by AMVs are outline in this section.

C.1. Oceanic Process Studies

AMVs used for process studies may be operating alone or in conjunction with research ships. Process studies may require the AMV to carry out measurements along pre-defined tracks, or alternatively, the AMV may be required to redefine its mission plan depending on results from its onboard sensors. In both cases, the mission plan for the AMV could also be modified via a communication link from a support vessel or a shore base. Typical missions might be:

    a. Study of the physical characteristics of Langmuir circulation cells near the ocean surface using upward-looking Doppler sonar (for currents), turbulence sensors, and sidescan sonar for imaging the associated bubble clouds; probably in association with a support ship with the

AMV pre-programmed to perform a track; mission duration likely to be less than one day

b. Study of decay processes within an oceanic mesoscale eddy using conductivity, temperature, and depth (CTD) seonsors, Doppler profilers, and turbulence sensors, possibly augmented with biological sensors

c. Mission involving onboard mission planning based on initial pre-programmed tracks, but modified in light of a feature map of the eddy under investigation compiled by the AMV itself; mission duration of several days, support vessel optional, could be a shore-based mission (see Section 4.2.3.)

d. Turbulence over sand banks, factors involved in creation and movement

e. Dynamics of boundary and bottom currents

f. Dispersion of pollutants from seabed sources

g. Biological responses to events, such as winter overturning in an anoxic basin

h. Dynamics of nutrient profiles in response to storm events.

C.2. Routine Observations

In this task, the AMV is more likely to be used to replace a surface vessel. The track will be pre-programmed, although small-scale, short-term deviations may arise from commands initiated by an onboard collision avoidance system. New mission plans, or an update to the existing plan, could be downloaded from a shore base via a communication channel, for example:

a. CTD profiles, with the vehicle travelling from the sea surface to near the seabed along a prescribed track, possibly deployed from a shore-base (either a coastal laboratory or a temporary location) as part of a long-term programme of observations

b. As above, but with the vehicle travelling between two moored or seabed docking stations, downloading data, recharging its energy supply and possibly updating its mission plan

c. Long-term seafloor ecosystem observation with biological and optical sensors along a standard track, either shore-based, ship-based, or between docking stations.

C.3. Survey

In the context of marine science research, AMVs may carry out reconnaissance or detailed surveys of the topography and characteristics of the seabed. Such surveys may use acoustic, optical, or chemical sensors. Tracks may either be pre-programmed (most likely for topographic surveys), with the proviso that the collision avoidance system may command short-term, small-scale variations, or the mission track may be dynamically altered by the mission management system. For example, the survey may be a:

a. Bathymetric and sidescan sonar topographic survey of a section of an ocean ridge, requiring the vehicle to terrain-follow at a pre-set altitude (typically 100m-plus) under bottom-track navigation

b. Optical survey of biological communities on the seafloor, requiring terrain-following at an altitude of typically 10m or less

c. Detection and mapping of buoyant plumes associated with hydrothermal vents, requiring onboard planning of a survey pattern in response to the results from chemical and/or optical sensors following an initial search pattern; altitudes may be 50–300m above the seabed.

C.4. Inspection

It is often difficult to distinguish between inspection and survey tasks, particularly as an AMV may be capable of carrying out both types and possibly in one mission. Inspection tasks usually concentrate on specific, pre-defined areas of offshore structures and subsea equipment. These tasks often will include detailed visual examination and other simple 'test' tasks of structures on the seabed or moored in the water column. This inspection data may be relayed back to the host facility in real time, or the inspection data/imagery will be recorded onboard the AMV for subsequent analysis in near real time.

Annex D: AMV Tools

D.1. Power Constraints

The range of useable AMV tools is currently constrained primarily by the available electrical power onboard the AMV. Users of AMVs have balanced (and will continue to balance) the power requirements range/diving activity against tool usage. This has meant that the AMV toolbox contains a wide range of 'passive', low-power-requirement environmental sampling devices. Examples of 'passive' tools are CTD or turbulence probes, which measure the properties of the water as it passes through the sensor area. Recently, there has been a growth of active tools, such as sonars and cameras covering a wider field of view, which have been designed to perform within the constraints of the available power on the AMV. New battery (and perhaps fuel cell) technologies will extend the range of tools available to the AMV community.

D.2. Marine Science AMV Tools

The tools carried by a marine science AMV will vary greatly in capability, size, weight, power consumption, and data storage requirements. While some AMVs may be dedicated to a set of tasks requiring standard sensor and tool sets, other vehicles will be used for a number of different purposes, with the need for adaptable and flexible mechanical and electrical systems to accommodate a wide variety of sensor and tool payloads.

D.3. *In Situ* Sensors

A large range of commercial off-the-shelf (COTS) sensors is available for use in AMVs covering the common physical, chemical, and biological measures. Such sensors may be self-contained, or may connect to the vehicle power and/or data distribution networks. In the latter case, consideration should be given to any effect on the vehicle data network performance and integrity. Techniques, such as transformer isolation, may be required for power and data.

These instruments will have differing requirements with regard to sensor exposure; in some cases a passive, ducted water flow will be most suitable, in other cases a pumped water flow will be needed, or the sensors may need to be exposed outside the body or fairing of the AMV. In each case, careful consideration will be needed to ensure that the data quality is not impaired and that any physical risk to the sensor or the people working on the vehicle is as low as practicable.

D.4. Sampling Devices

Notwithstanding rapid advances in *in situ* sensor technology suitable for use on AMVs, there remains a need for sampling devices to gather water samples either for calibration (e.g. for salinity or nutrient concentration) of those substances that cannot yet be measured *in situ* (e.g. chlorofluorocarbons). Careful consideration will need to be given to siting the intakes for such sampling devices to avoid trace element contamination from the vehicle structure and to ensure safe handling when the containers are returned to the surface. Samplers may also be used to capture both phytoplankton and zooplankton for later analysis. Such samplers may incorporate chemical preservatives such as ethyl alcohol or formalin; the use of such substances and any potential for spillage, leakage, or pollution should be covered in the mission risk assessment.

Marine science AMVs with the ability to hover may also be used to physically sample seabed sediments using coring tools. AMVs used for these purposes will include buoyancy change compensation mechanisms.

D.5. Remote Sensing Devices

The capability of marine science AMVs can be greatly augmented through adding remote sensing instruments for water column or seabed studies.

Remote sensing instruments will use sonar for long-range sensing or light at shorter ranges. In general, the power requirement of remote-sensing instruments is higher than that of *in situ* instruments, and the vehicle energy budget and power supply capacity should be carefully planned to account for both the peak and continuous power requirements of these systems.

Consideration should be given to the potential for mutual interference between remote sensing sonars (e.g. sidescan, bathymetric, Doppler sonars, and sub-bottom profilers) and the vehicle's standard acoustic systems used for navigation, telemetry, and location.

D.6. Industrial AMV Tools

The applications of AMV for industrial and commercial purposes currently are focused on investigation of seabed topography and composition, as well as inspection of man-made structures. For these types of measurements, various types of active sonar are the most appropriate, together with optical cameras where water clarity permits their use.

To gain the best quality data products from these sensors requires a high degree of navigational accuracy, relative to the seabed being studied. This is generally achieved using a combination of systems, and in contrast to the research role, an acoustic-tracking system is predominantly used and usually requires that the host vessel remains close to the AMV throughout its mission.

Otherwise, the range and types of AMV tools available for use during an industrial/commercial mission will be identical to those available for a research mission.

D.7. Military AMV Tools

Military AMV tools will embrace many of the applications mentioned above. The nature of the application of AMVs to military problems indicates there are likely to be additional highly specialised items in the military AMV toolbox.

ANNEX E: MEMBERSHIP OF THE SUT AUTONOMOUS UNDERWATER VEHICLE LEGAL WORKING GROUP (AUVLWG) 2009

Members of the AUVLWG contributing to the updating of the original *The Operation of Autonomous Underwater Vehicles, Volume One: Recommended Code of Practice*, First Edition (2000) are:

Keith Broughton – Chairman	The Leviathan Facility
Chris Ashcroft	International Marine Solutions
Andrew Chamberlain	Holman Fenwick Willan
Captain Ralph Coton	The Shipowners' Protection Ltd
Alex Davis	Stephenson Harwood
Ian Gallett*	Society for Underwater Technology
Prof Gwyn Griffiths*	National Oceanographic Centre, Southampton
Jim Jamieson	Subsea7*
Tony Marshall	Kongsberg Maritime
Trevor Newman*	Dera
Stephen Phillips	Autonomous Surface Vehicles Ltd
Alan Thomas	Charles Taylor Consulting

---

* Indicates those AUVLWG members who are also involved in the SUT Underwater Robotics Group

Attributions reflect when the member was most active in the group. They are intended to indicate the breadth of experience of the group and in no way indicate endorsement by the member's company of document.

The members are very grateful to all those others who have commented and attended the workshops. They are also very grateful to the original team who produced volume one in 2000 (attributes as originally shown, although biographical details as originally shown omitted):

Dr J Dering* – Chairman	Defence Evaluation and Research Agency Haslar
Mr C Carleton MBE	Head of the Law of the Sea Department, Hydrographic Office
Dr C Fay*	Director, Research Vessel Service, Southampton Oceanography Centre
Mr G Griffiths	Head of Ocean Technology Division, Southampton Oceanographic Centre (a joint venture between the Natural Environment Research Council and the University of Southampton)
Lt Cdr A Holt, Royal Navy	Directorate of Naval Surveying, Oceanography and Meteorology
Lt Cdr R J Rogers, Royal Navy	Naval Liaison Officer, Defence Science and Technology Laboratories, Winfrith
Mr A Tonge*	Project Manager, BAE Systems

---

* Indicates those AUVLWG members who are also involved in the SUT Underwater Robotics Group

APPENDIX 1: STATUS OF THE LAW/CURRENT LEGAL ISSUES REGARDING AMVS

A. The Law as Contained in the Report Published by the Society for Underwater Technology in January 2000

1. Application of the Merchant Shipping Act 1995 to AMVs

1.1. In order to see if any or all of the provisions of the Merchant Shipping Act apply to AMVs (as a form of ship), it is necessary to consider the general principles laid down in the case law as to the application of the Merchant Shipping Act definition of 'ship' and see how far they would apply by analogy to AMVs. It will then be necessary to continue by examining each particular circumstance (salvage, registration, etc.) in order to see if any special definition is applicable, so that for some purposes an AMV is to be treated as a ship.

1.2. S313(1) of the Merchant Shipping Act 1995 provides that 'unless the context otherwise requires … "ship" includes every description of vessel used in navigation'.

1.3. Case law considering whether a vessel can be considered a 'ship' has looked to its means of propulsion, its area of work, and the object of its work. The case *Steedman v. Scofield* (1992, 2 Lloyd's Rep 163) emphasised the need for a vessel to be 'used in navigation' for it to be considered a 'ship' under the Merchant Shipping Act.

Although this decision been criticised and does not fully rule out the possibility of an AMV being a ship, it does give the impression that a judge may be unlikely to consider it as such. Since there are such a wide variety of AMVs used for very different purposes, it is possible that some may be recognised as ships, whereas others may not.

1.4. It appears from the provisions of the Merchant Shipping Act 1995 itself that its regulatory regime for ships will not generally apply to AMVs unless the context indicates otherwise. Section 88 of the Act was specifically designed to regulate manned submersibles, inferring that submersibles were not included in the definition of a ship, but ignoring the question of whether the Act would apply to AMVs.

1.5. Section 311 of the Merchant Shipping Act 1995, since repealed and amended by Section 112 of the Railways and Transport Safety Act 2003, recognised the potential definitional problems with 'ship' by giving the secretary of state power to provide that certain structures designed or adapted for use at sea are to be treated as ships. Although these powers have not been exercised, there is no doubt that they could be exercised to apply merchant shipping legislation to AMVs. The question is whether the legislation as a whole should be extended to apply to AMVs, or whether it should be decided to what extent each provision should apply to AMVs.

1.6. There is the potential for delegated legislation made under the Merchant Shipping Act which extends to application to AMVs. So far there are no regulations which have been extended explicitly to unmanned submersibles, however, there are varying definitions of 'ship' contained in existing delegated legislation. For example, the regulations implanting Safety of Life at Sea (SOLAS) on the whole is directed to seagoing cargo ships, whereas the regulations giving effect to International Convention for the Prevention of Pollution from Ships (MARPOL) give a definition of ship extending to 'submersible craft'.

2. *Application of Current International Maritime Conventions to AMVs*

2.1. Each international convention describes the scope of its own application, which will have to be interpreted looking at, amongst other things, the object and purpose of the convention. Many international maritime conventions are stated to apply to 'ships'. There is, however, no uniform definition of 'ship' in the various international maritime conventions that can be used to decide whether those conventions apply to AMVs. Whether a particular convention can be applied to AMVs depends on the definition within (and therefore scope of) of that particular convention.

2.2. It is further not clear how these definitions will be interpreted when incorporated in national law.

> 2.2.1. *SOLAS 1974/1978/1988 and the International Management Code for the Safe Operation of Ships and for Pollution Prevention (ISM Code) 1994*
> The SOLAS Convention states that it shall apply to 'ships entitled to fly the flag of State the Governments of which are contracting Governments'. As the convention gives a very particular list of categories of ship, none of which would easily cover the present generation of AMVs, it may be concluded that its provisions should not apply generally to AMVs.
>
> 2.2.2. *Load Lines Convention 1966*
> This convention is clearly not designed for submersibles such as AMVs.
>
> 2.2.3. *Tonnage Measurement Convention 1969*
> It appears that the present generation of AMVs would not be covered by this convention, either because they are not ships within its definition, or are not long enough.
>
> 2.2.4. *Convention on the International Regulations for Preventing Collisions at Sea (COLREG) 1972*
> This applies to 'all vessels upon the high seas'. Although the definition of vessel includes the phrase 'used as a means of transportation on water', it is arguable that an AMV should be included within the definition of a vessel due to the breadth of the categories of craft to which the convention appears to be designed to apply, although the rules only apply to surface navigation.

2.2.5. MARPOL 1973/1978

This convention specifically includes submersibles within the definition of 'ship', which would seem to cover AMVs, although it is possible to argue that submersibles still have to be a type of vessel, therefore excluding unmanned craft. MARPOL is, in addition, stated to apply to 'ships entitled to fly the flag' of a state party; there is a question as to whether AMVs can be flagged at all.

2.2.6. *The Convention on the Prevention of Marine Pollution by Dumping of Wastes and Other Matter (LDC) 1972/1996*

It is unclear whether this convention applies to AMVs. It is stated to apply to waterborne craft, which may or may not include submersibles, however, it would not have a great impact since it is not the purpose of an AMV to dump material. (Contrast that MARPOL is specifically stated to apply to submersibles.)

2.2.7. *Convention for the Protection of the Marine Environment of the North East Atlantic (OSPAR) 1992*

The definition of vessels to which this convention applies includes 'other man made structures in the maritime area', which could be seen to extend to AMVs, although again the definition does not specifically include submersibles.

2.2.8. *International Convention on Standards of Training, Certification, and Watchkeeping for Seafarers (STCW Convention) 1978/1995*

The critical question here is whether an AMV is a ship to which this convention applies; if it is considered to be such, then its operation on the surface without any watchkeeping arrangements might be an offence.

2.2.9. *Memorandum of Understanding on Port State Control (Paris MOU) 1982*

If the instruments to which the Paris MOU is related (SOLAS, MARPOL, STCW, COLREG, and the Tonnage Measurement Convention 1969) are found to apply to AMVs, then the MOU will apply to its enforcement.

2.2.10. *Suppression of Unlawful Acts Convention 1988*
This is stated to apply to submersibles and may specifically apply to AMVs, since the unlawful acts it covers are those said to jeopardise the safety of persons and property.

2.2.11. *Intervention Convention 1969/1973*
The broad definition of ship under this convention may be wide enough to cover AMVs, although the definition under the implementing UK legislation suggests that AMVs are not covered.

2.2.12. *International Convention on Civil Liability for Oil Pollution Damage (CLC) 1992 and Fund Convention 1992*
The definition of ships to which these conventions apply clearly excludes the present generation of AMVs, which are not constructed or adapted for the carriage of oil in bulk as cargo.

2.2.13. *International Convention on Liability and Compensation for Damage in Connection with the Carriage of Hazardous and Noxious Substances by Sea (HNS Convention) 1996*
The definition of ship to which this applies includes 'seaborne craft, of any type whatsoever', which would seem to cover an AMV, although the convention applies to pollution from 'hazardous and noxious substances' which are 'any substances, materials and articles carried on board a ship as cargo'. It is unlikely that an AMV would ever have a cargo other than merely equipment designed for use at sea.

2.2.14. *International Convention on Civil Liability for Bunker Oil Pollution Damage 2001*
The convention entered force on 21 November 2008, however, since AMVs are electrically driven and do not carry bunker oil, it will not be of relevance.

2.2.15. *The Nairobi Wreck Removal Convention 2007*
The convention was agreed at a conference between 14 to 18 May 2007 and since then four countries have signed the convention: Netherlands, Estonia, France and Italy. The convention will be formally ratified when ten countries

have signed the convention. Within the convention, a 'ship' is defined as 'a vessel of any type whatsoever operating in the marine environment and includes ... submersibles'. 'Wreck' means 'a sunken or stranded ship, or any part thereof, including anything that is or has been onboard such a ship'. The definitions therefore appear to be wide enough to include AMVs.

2.2.16. Collision Convention 1910

This convention sets out a two-year time bar for claims resulting from collisions. The convention applies to collisions between 'seagoing vessels or between seagoing vessels and vessels capable of inland navigation'. The convention assumes that there will be a master onboard the vessel, and it seems unlikely that the convention will apply to AMVs, e.g. if they are in collision with a ship. This means that national rules on the apportionment of liability will apply, as will general national time bars, which in the UK are longer than two years.

2.2.17. Salvage Convention 1989

Under this convention, salvage operation means any act or activity undertaken to assist 'a vessel or any other property in danger'. This convention may be relevant where an AMV is used to salve other property, as well as where it is itself salved. Vessel is defined as 'any ship or craft, or any structure capable of navigation'. This appears to be broad enough to cover AMVs so long as they are found to be 'capable of navigation'. In any case, an AMV would fall within the definition of 'any other property'.

2.2.18. Convention on Limitation of Liability for Maritime Claims (LLMC) 1976 and 1996 Protocol

See Section 7.2 of this appendix.

2.2.19. Ship Registration Convention 1986

The convention defines 'ship' to mean 'any self propelled seagoing vessel used in international seaborne trade for the transport of goods, passengers, or both with the exception of vessels of less than 500 gross registered tonnes'. AMVs would not come within this definition.

2.2.20. Maritime Liens and Mortgages Conventions 1926/1967/1993
It is doubtful that these conventions were designed to apply to AMVs or if they will be of much relevance.

2.2.21. Arrest Conventions 1952/1999
These will only apply if AMV is considered to be a 'ship', which is not defined in either convention. It seems unlikely that they will apply to an AMV.

2.2.22. Hague Rules 1924, Hague-Visby Rules 1968, Hamburg Rules 1978
These apply to 'goods' carried on a 'ship', which is defined as 'any vessel used for the carriage of goods by sea'. Even if an AMV did carry cargo, it is unlikely that carriage documentation under these conventions would be issued in respect of it. In the unlikely event that carriage in an AMV did occur, it would be likely to qualify under Article 6 of the Hague Visby Rules, which allows a non-negotiable receipt to contain any terms whatsoever. The carriage conventions could be relevant where an AMV is itself carried as cargo in a ship.

3. Development of Specific AMV Regimes: Draft ODAS Convention 1993

This draft convention was created to set out a regime for 'offshore data acquisition systems' and would have dealt specifically with the legal issues relating to AMVs.

This draft convention would include requiring states to establish a special register system for ODAS, provided that any ODAS is not subject to a vessel registry system (as it would not generally be under the Ship Registration Convention 1986).

4. Property Issues and AMVs

4.1. Registration and Mortgages

Ships are usually registered, aligning themselves with a flag state which will provide protection as well as imposing various regulations.

Registration enables finance to be obtained, as the lender can register its mortgage and know that it will be obtain priority over other creditors. The mortgagee is also given rights to enforce its security. In order to be registered in the UK as a ship, an AMV would have to meet the definition of ship in S313(1) of the Merchant Shipping Act 1995. It is likely, however, that it would not satisfy this requirement.

There is the potential that maritime liens may exist in relation to an AMV. This again depends on whether or not an AMV is classed as a ship or vessel.

4.2. Enforcement of Claims – Civil Detention

The Arrest Conventions 1952 and 1999 will only apply to an AMV if it is a ship. Many national laws, however, allow for the seizure or attachment of property in order to support civil claims. Since the restrictions on arrest in these conventions would not apply, it would be down to the national rules to determine whether an AMV could be detained.

5. *Liability and Compensation: Non-Contractual Liabilities*

5.1. There are various legal basis of non-contractual liability – including negligence, strict liability regimes (such as that under the Harbours, Docks and Piers Clauses Act 1847) and salvage claims – that will apply to incidents involving AMVs leading to damage to property and/or persons.

5.2. Even though the maritime standards set out in COLREG, the International Maritime Dangerous Goods (IMDG) Code, or the ISM Code might not strictly apply to AMVs, the courts would probably measure the standard of care to be expected from AMV owners against these standards. The AMV CoP may also become significant in assessing any negligence.

6. *Liability and Compensation: Contractual Liability*

An AMV operator could enter into many different types of contract, including demise/bareboat charters, time charters, voyage charter, research services contracts, hire of the AMV(with or without team), etc. The basic rules of the law of contract would apply to these contracts, although the contracts themselves would have to be tailored to meet the particular needs of the parties.

7. Limitation of Maritime Liabilities

This is a critical area for those owning and operating AMVs.

7.1. Potential Use of Limited Company to Shield from Unlimited Liability

Limited liability companies have traditionally been used to keep any liability accrued by a company within that company. This is not, however, a solve-all solution. Where a company only has limited resources (such as companies established by universities to develop scientific innovations), any contractor contracting with that company would want security (e.g. a guarantee) from outside that company (e.g. from the research institute) or insurance.

There are occasions where the corporate veil of the company can be broken and those behind the company held liable for sums the company does not have. In addition, it may be found that people outside the limited company are responsible in their own right. This makes the limitation of liability by law a very desirable solution.

7.2. International Conventions on Limitation of Liability

The LLMC 1976 replaced the International Convention Relating to the Limitation of the Liability of Owners of Seagoing Ships, which was signed in Brussels in 1957, and came into force in 1968 (and is still in force in some states).

The right to limit is a privilege granted to owners, charterers, managers, or operators of 'ships'. The LLMC does not contain a separate definition of 'ship', so it is left to national interpretation.

Where ships are insured, the existence of the principle of limitation of liability enables insurance cover to be offered at lower rates than would otherwise be available, as the insurer is effectively able to take the benefits of the limits on behalf of the assured.

This makes it important to establish whether an AMV is covered by the definition of 'ship' in the existing limitation conventions and implementing national legislation. Only then will it be necessary to look further at those entitled to limit and the amount of the limit.

The LLMC 1976 was enacted in the UK in the Merchant Shipping Act 1995, Schedule 7. The Merchant Shipping (Convention on

Limitation of Liability for Maritime Claims) (Amendment) Order 1998 amended Schedule 7 to implement the 1996 Protocol when it came into force.

Article 1(1) of the LLMC 1976 states that 'shipowners and salvors, as hereinafter defined, may limit their liability in accordance with the rules of this Convention'. Article 2(2) provides that the 'term "shipowner" shall mean the owner, charterer, manager or operator of a sea-going ship'. Ship is not otherwise defined under the convention.

For the purposes of English law, the Merchant Shipping Act 1997, Schedule 7, Part II, paragraph 2, provides that the 'right to limit liability under the Convention shall apply in relation to any ship whether seagoing or not, and the definition of "shipowner" in paragraph 2 of article 1 shall be construed accordingly'.

Paragraph 12 on the meaning of ship states that 'references in the Convention and in the preceding provisions of this Part of this Schedule to a ship include references to any structure (whether completed or in course of completion) launched and intended for use in navigation as a ship or part of a ship'.

This again means that the question has to be asked as to whether an AMV comes within the definition of a ship under English law.

Even if an AMV is not found to come within the definition of a ship, if it is being operated from a mother ship, it may be possible to argue that the liability arises out of the operation of the mother ship, so the limits of that ship would apply. Any argument along these lines would very much depend on the facts of each particular case*.

8. Insurance Cover

8.1. There is no international convention which requires an AMV to carry insurance cover. The extent to which insurance cover is required, if at all, will vary according to national law.

---

* Where a state is involved in an AMV project, there is the potential for the application of state immunity, although this is generally restricted where the state is involved in commercial operations. Note: even if an AMV was covered under a convention there may be a complication with determining the amount of the limit as it appears that no AMV at present will have a registered tonnage (the basis for the calculation of the limit).

Non-binding guidelines on the provision of financial security were approved by the 80th Session of the IMO Legal Committee in October 1999 and were submitted to the IMO Assembly as a resolution. This led to Resolution A.898(21), Guidelines on Shipowners' Responsibilities in Respect of Maritime Claims, being adopted on 25 November 1999. It provides that ship owners should insure themselves for claims for which they are able to limit their liabilities. This is guidance only, but may result in states applying national law to force ship owners to carry certain forms of insurance.* These guidelines are contained in the Maritime and Coastguard Agency's Marine Guidance Note, MGN 135 (M).

8.2. Section 192A Merchant Shipping Act 1995 gives the Secretary of State power to make regulations requiring that there must be a contract of insurance in place for a ship while that ship is in UK waters. (No such regulations have been made, and in any case it is unlikely that they would apply to AMVs unless the definition of its application was broadened beyond 'ship'.)

8.3. The appropriate marine insurance for an AMV would be the same as that used for ships, namely for 'hull and machinery' (cover for damage to property) and 'liability' (cover from claims by third parties). It is necessary to obtain specialist advice from marine insurers on the insurance that should be taken, due to the specialist nature of AMV operations.

8.4. If AMV operators, funding bodies, etc., are using 'hold harmless' clauses to remove rights to claims between the parties, this may have an impact on any insurance policy or their ability to find suitable insurance.

B. Changes in the Law since the Publication of the Report in January 2000 until to June 2009

1. *Domestic Law*

1.1. In the case *R v. Goodwin* (2005, EWCA Crim 3184; ; 1 Lloyds Rep 432), the Court of Appeal had to decide whether a jet ski was a

---

* Ship owners are urged to comply with these guidelines in respect of all seagoing ships of at least 300 gross tonnage, while are also encouraged to comply in respect of ships of less than 300 gross tonnage. There is again the question of whether this was intended to apply to owners of AMVs, since it is based on the existing limitation of liability regimes.

ship. In doing this, it considered the construction of the vessel and the term 'used in navigation'. The Court of Appeal found that for a vessel to be 'used in navigation' under the Merchant Shipping Acts, it is not a necessary requirement that it should be used in transporting persons or property by water to an intended destination, distinguishing *Steedman v. Scofield*. It considered that what was critical was whether, for the purposes of the Merchant Shipping Act definition of ship, navigation was 'the planned or ordered movement from one place to another', or whether it could extend to 'messing about in boats' involving no journey at all. The court concluded that the correct authorities were those that confined 'vessel used in navigation' to 'vessels which are used to make ordered progression over the water from one place to another'. This would, however, still seem to confirm the position that an AMV would not be considered a ship for the purposes of the Merchant Shipping Act.

1.2. There have been no amendments to the Merchant Shipping Act 1995, or new regulations under it, or any other domestic legislation which change the 2000 legal analysis.

2. International Law

2.1. Development of Specific AMV Regime: Draft ODAS Convention 1993

There does not appear to have been any progress with this draft convention since 2000.

2.2. Limitation of Liability

The LLMC 1996 entered into force on 13 May 2004 for the states party to it. This protocol does not, however, extend the scope of application of the convention.

2.3. International Conventions

2.3.1. Load Lines Convention 1966

The 1988 Protocol entered into force 3 February 2000, and the 2003 amendments entered into force on 1 January 2005. These changes have not altered the fact that it was not designed for submersibles such as AMVs.

2.3.2. Maritime Liens and Mortgages Conventions 1926/1967/1993

The International Convention on Maritime Liens and

Mortgages 1993 entered into force on 5 September 2004.

It is doubtful that these conventions were designed to apply to AMVs, or if they will be of much relevance.

2.3.2. *The UNIDROIT Convention on International Interests in Mobile Equipment 2001*

The International Institute for the Unification of Private Law (UNIDROIT) Convention on International Interests in Mobile Equipment (November 2001) entered into force on 1 April 2004 and harmonises the laws of secured transactions where the collateral consists of mobile equipment. Signatories include the United States, France, Germany, and the United Kingdom. This might possibly have some relevance to the commercial financing of AMVs, but not in the immediate future.

C. Salvage

1. Incidents leading to insurance claims through an AMV causing damage to third-party property or injury to persons are expected to be rare. With their unmanned status, however, AMVs present a number of interesting issues with regard to salvage law. The limited (but growing) experience of AMV operations available to date suggests two main risk areas: first, damage to vehicle during deployment or recovery to/from the mother platform; and second, vehicle loss during mission, which usually proves to be temporary with the vehicle later being located (although with varying degrees of damage).

2. With regard to vehicle recovery/deployment, under the draft definition of an AMV (as outlined in the main body of the briefing document), a vehicle which is still mechanically attached to the mother vessel (e.g. by a crane) is not an AMV. Indeed, during deployment operations, the risks to the vehicle are not in themselves AMV risks. Instead, they are generic risks relevant to all marine overside equipment when in the vicinity of the deployment vessel.

3. The most likely day-to-day risk with AMV operations would be temporary loss of the vehicle. In the case of coastal operations in particular, it is probable that the vehicle will be found by a third party, and thus that the

issue of salvage will arise. Salvage case law would appear to be clear in such an instance, although recent incidents concerning the Royal Navy overside equipment found and recovered by other water users (PAP vehicle being re-covered/taken by a fisherman) must give an AMV operator some concern.

4. AMV operators and their insurers would therefore benefit from clarity on the legal aspects of AMV salvage issues that would to be applicable to all water users. Any such clarification should also seek to deal with unnecessary salvage operations (either malicious or with good intent) on an AMV when it is undergoing its normal operation and when, although not in immediate contact with the operator, it is not in any danger.

5. The following key points stand out from this discussion:

    - An AMV needs to be protected from malicious interference and/or unintended interference (i.e. a 'salvor' recovering a correctly functioning/operating unit which is not in distress)

    - This may require the formulation of a method of allowing third parties to know when a unit is in distress and requires assistance

    - Agreed principles of AMV salvage should made available to all water users. It could be considered whether it is possible to formulate a standard method of indicating how to make the vehicle safe, including, for example, how to switch it off, where to attach lift points, how heavy it is, presence of hazardous materials, etc.

Appendix 2: IMO – Revised Technical Annex II to the Draft Convention on the Legal Status of Ocean Data Acquisition Systems (ODAS)*

Noting that the Intergovernmental Oceanographic Commission and the World Meteorological Organization had concurred with a proposed revised Technical Annex II to the draft Convention on the Legal Status of ODASs, the Maritime Safety Committee adopted at its 49th session the revised text, which is attached.

1. Identification and Marking

1.1. General

1.1.1. Every ODAS shall be assigned a unique identification number prefixed by the letters 'ODAS' and suffixed by letters indicating the state in which each is registered, in full or in a commonly abbreviated form.

1.1.2. Each ODAS shall display its alpha numerical identification assigned in accordance with paragraph 1.1.1. clearly on an exterior surface where it can best be seen and, in addition, if feasible, the name and address of its owner.

1.1.3. A replica of the flag of the state in which the ODAS is registered may also be painted on or applied to the exterior surface as a further optional means of identification.

---

* (MSC/Circ.372, 14 June 1984)

1.2. Surface-Penetrating ODAS

Surface-penetrating ODAS shall have its visible positions painted yellow. Drifting ODAS should carry an inscription in several languages stating that its purpose is to drift freely and that it should not be recovered by unauthorised persons.

2. Lights and Signals

2.1. General

2.1.1. The lights and signals referred to hereunder shall be positioned in places where they can best be seen or heard.

2.1.2. A satisfactory radar response at a distance of at least 2 miles shall be ensured for an ODAS which constitutes a danger to shipping and safe navigation, and effort shall be made to increase this range where the size of the ODAS allows.

2.2. Surface-Penetrating ODAS of All Types Other than Bottom-Bearing ODAS

2.2.1. Surface-penetrating ODAS of all types other than bottom-bearing ODAS shall:

    a. exhibit from sunset to sunrise and, in the case of manned ODAS, also in conditions of poor visibility, a yellow light visible all round the horizon with, where technically practicable, a nominal range of at least 5 miles, exhibiting a group of 5 flashes every 20 seconds, the flash rate not to exceed 30 per minute; and

    b. carry a sound signal, where the installation thereof is technically practicable, of such a nature that it cannot be confused with neighbouring aids to navigation, nor with sound signals made in compliance with the International Regulations for Preventing Collisions at Sea.

2.3. Surface-Penetrating ODAS which Are Bottom-Bearing ODAS

2.3.1. Bottom-bearing, surface-penetrating ODAS shall be marked and carry lights and sound signals in the same manner as 'a structure in the sea', e.g. drilling platforms, as is customary in the area concerned.

2.4. Sub-Surface ODAS

2.4.1. Sub-surface ODAS of all types that, due to the depth at which they are deployed, constitute a danger to shipping and safe navigation or fishing gear shall, when they are not escorted by an attending vessel capable of giving warning(s) of their presence to passing ships, be marked by a surface buoy exhibiting lights and complying with the requirements for sound signals in paragraph 2.2.1.

3. Modification or Waiver

3.1. The requirements of 1 and 2 may be modified or waived by the registry state subject, where relevant, to the concurrence of the state providing aids to navigation in the area concerned and at the risk of the operator, if such a waiver or modification does not result in the ODAS becoming a danger to shipping and safe navigation.